ENERGY SECTOR STANDARD OF THE PEOPLE'S REPUBLIC OF CHINA

中华人民共和国能源行业标准

Specification for Engineering Geological Investigation of Underground Structures for Hydropower Projects

水电工程地下建筑物工程地质勘察规程

NB/T 10241-2019

Replace DL/T 5415-2009

Chief Development Department: China Renewable Energy Engineering Institute
Approval Department: National Energy Administration of the People's Republic of China
Implementation Date: May 1, 2020

China Water & Power Press

Beijing 2024

All rights reserved. No part of this publication may be reproduced, stored in a retrieval system, or transmitted in any form or by any means—electronic, mechanical, photocopying, recording or otherwise, without prior written permission of the publisher.

图书在版编目（CIP）数据

水电工程地下建筑物工程地质勘察规程：NB/T 10241-2019 Replace DL/T 5415-2009 = Specification for Engineering Geological Investigation of Underground Structures for Hydropower Projects (NB/T 10241-2019 Replace DL/T 5415-2009)：英文 / 国家能源局发布. -- 北京：中国水利水电出版社，2024. 7. -- ISBN 978-7-5226-2635-2

Ⅰ.P642-65

中国国家版本馆CIP数据核字第2024FR7744号

ENERGY SECTOR STANDARD
OF THE PEOPLE'S REPUBLIC OF CHINA
中华人民共和国能源行业标准

Specification for Engineering Geological Investigation
of Underground Structures for Hydropower Projects
水电工程地下建筑物工程地质勘察规程
NB/T 10241-2019
Replace DL/T 5415-2009
（英文版）

Issued by National Energy Administration of the People's Republic of China
国家能源局　发布
Translation organized by China Renewable Energy Engineering Institute
水电水利规划设计总院　组织翻译
Published by China Water & Power Press
中国水利水电出版社　出版发行
　　Tel: (+ 86 10) 68545888　68545874
　　sales@mwr.gov.cn
　　Account name: China Water & Power Press
　　Address: No.1, Yuyuantan Nanlu, Haidian District, Beijing 100038, China
　　http://www.waterpub.com.cn
中国水利水电出版社微机排版中心　排版
北京中献拓方科技发展有限公司　印刷
184mm×260mm　16开本　3.5印张　111千字
2024年7月第1版　2024年7月第1次印刷

Price（定价）：￥580.00

Introduction

This English version is one of China's energy sector standard series in English. Its translation was organized by China Renewable Energy Engineering Institute authorized by National Energy Administration of the People's Republic of China in compliance with relevant procedures and stipulations. This English version was issued by National Energy Administration of the People's Republic of China in Announcement [2023] No. 8 dated December 28, 2023.

This version was translated from the Chinese Standard NB/T 10241-2019, *Specification for Engineering Geological Investigation of Underground Structures for Hydropower Projects*, published by China Water & Power Press. The copyright is reserved by National Energy Administration of the People's Republic of China. In the event of any discrepancy in the implementation, the Chinese version shall prevail.

Many thanks go to the staff from the relevant standard development organizations and those who have provided generous assistance in the translation and review process.

For further improvement of the English version, any comments and suggestions are welcome and should be addressed to:

China Renewable Energy Engineering Institute
No. 2 Beixiaojie, Liupukang, Xicheng District, Beijing 100120, China
Website: www.creei.cn

Translating organization:

POWERCHINA Chengdu Engineering Corporation Limited

Translating staff:

ZHANG Yixi	ZHAO Cheng	ZHOU Yifei	DING Zihan
CHEN Weidong	WU Zhanglei	Li Qingchun	DENG Zhongwen
LI Jianming	HE Lei	DU Xiaoxiang	HU Yadong
FAN Guijun	JIANG Tingting		

Review panel members:

GUO Jie	POWERCHINA Beijing Engineering Corporation Limited
LIU Qing	POWERCHINA Northwest Engineering Corporation Limited

LI Zhongjie	POWERCHINA Northwest Engineering Corporation Limited
ZHANG Ming	Tsinghua University
PENG Peng	POWERCHINA Huadong Engineering Corporation Limited
ZHAO Daiyao	POWERCHINA Guiyang Engineering Corporation Limited
CHEN Li	POWERCHINA Kunming Engineering Corporation Limited
GAO Yan	POWERCHINA Beijing Engineering Corporation Limited
ZHANG Dongsheng	China Renewable Energy Engineering Institute
ZHANG Weiheng	China Renewable Energy Engineering Institute

National Energy Administration of the People's Republic of China

翻译出版说明

本译本为国家能源局委托水电水利规划设计总院按照有关程序和规定，统一组织翻译的能源行业标准英文版系列译本之一。2023年12月28日，国家能源局以2023年第8号公告予以公布。

本译本是根据中国水利水电出版社出版的《水电工程地下建筑物工程地质勘察规程》NB/T 10241—2019翻译的，著作权归国家能源局所有。在使用过程中，如出现异议，以中文版为准。

本译本在翻译和审核过程中，本标准编制单位及编制组有关成员给予了积极协助。

为不断提高本译本的质量，欢迎使用者提出意见和建议，并反馈给水电水利规划设计总院。

地址：北京市西城区六铺炕北小街2号
邮编：100120
网址：www.creei.cn

本译本翻译单位：中国电建集团成都勘测设计研究院有限公司
本译本翻译人员：张一希　赵　程　周逸飞　丁梓涵
　　　　　　　　陈卫东　吴章雷　李青春　邓忠文
　　　　　　　　李建明　贺　磊　杜潇翔　胡亚东
　　　　　　　　樊贵军　蒋婷婷

本译本审核成员：
郭　洁　中国电建集团北京勘测设计研究院有限公司
柳　青　中国电建集团西北勘测设计研究院有限公司
李仲杰　中国电建集团西北勘测设计研究院有限公司
张　明　清华大学
彭　鹏　中国电建集团华东勘测设计研究院有限公司
赵代尧　中国电建集团贵阳勘测设计研究院有限公司
陈　砺　中国电建集团昆明勘测设计研究院有限公司
高　燕　中国电建集团北京勘测设计研究院有限公司

张东升　水电水利规划设计总院
张伟恒　水电水利规划设计总院

国家能源局

Announcement of National Energy Administration of the People's Republic of China [2019] No. 6

National Energy Administration of the People's Republic of China has approved and issued 384 energy sector standards including *Technical Specification for Electrical Exploration of Hydropower Projects* (Attachment 1), the English version of 48 energy sector standards including *Technical Guide for Rock-Filled Concrete Dams* (Attachment 2), and Amendment Notification No. 1 for 7 energy sector standards including *Technical Code for Environmental Impact Assessment of Wind Farm Projects* (Attachment 3); and abolished 5 energy sector standards/plans including *Charging Standards for Investigation and Design of Wind Power Project* (Attachment 4).

Attachments: 1. Directory of Sector Standards

2. Directory of English Version of Sector Standards

3. Amendment Notification for Sector Standards

4. Directory of Abolished Sector Standards/Plans

National Energy Administration of the People's Republic of China

November 4, 2019

Attachment 1:

Directory of Sector Standards

Serial number	Standard No.	Title	Replaced standard No.	Adopted international standard No.	Approval date	Implementation date
…						
18	NB/T 10241-2019	Specification for Engineering Geological Investigation of Underground Structures for Hydropower Projects	DL/T 5415-2009		2019-11-04	2020-05-01
…						

Foreword

According to the requirements of Document GNKJ [2015] No. 283 issued by National Energy Administration of the People's Republic of China, "Notice on Releasing the Development and Revision Plan of Energy Sector Standards in 2015", and after extensive investigation and research, summarization of practical experience, and wide solicitation of opinions, the drafting group has prepared this specification.

The main technical contents of this specification include: basic requirements, engineering geological investigation content of underground structures, engineering geological investigation methods for underground structures, engineering geological assessment of surrounding rock of underground structures, and engineering geological work during underground structure construction.

The main technical contents revised are as follows:

— Adding the content of environmental protection in investigation.

— Adding the content of investigation on ground temperature, water inrush, and mud inrush.

— Adding the content of mud inrush prediction of fault-fractured zone with a large amount of mud.

— Adding the content of assessment on heat harm of ground temperature.

— Adding the content of engineering geological assessment on stability of surrounding rock for air cushion surge chambers.

— Adjusting the content of the engineering geological investigation during underground structure construction, simplifying "geological work for surrounding rock treatment" and "geological work for surrounding rock monitoring", and integrating the content into "geological work during construction".

— Adjusting the content of classification of gas tunnels.

National Energy Administration of the People's Republic of China is in charge of the administration of this specification. China Renewable Energy Engineering Institute has proposed this specification and is responsible for its routine management. Energy Sector Standardization Technical Committee on Hydropower Investigation and Design is responsible for the explanation of specific technical contents. Comments and suggestions in the implementation of this specification should be addressed to:

China Renewable Energy Engineering Institute

No. 2 Beixiaojie, Liupukang, Xicheng District, Beijing 100120, China

Chief development organization:

POWERCHINA Chengdu Engineering Corporation Limited

Chief drafting staff:

LI Wengang	CHEN Weidong	LIAO Mingliang	YANG Jianhong
DENG Zhongwen	LI Panfeng	ZHANG Yunda	SHA Chun
ZHANG Yong	JU Lin		

Review panel members:

YANG Jian	WANG Zongli	ZHANG Dongsheng	GUO Yihua
MI Yingzhong	GONG Hailing	SHAN Zhigang	CHEN Liqiang
LU Dongliang	ZENG Xiangxi	ZHANG Guofu	LI Kaide
GAO Qiang	WU Yongfeng	DAI Qixiang	WANG Huiming
LI Shisheng			

Contents

1	General Provisions	1
2	Terms	2
3	Basic Requirements	3
4	**Engineering Geological Investigation Content of Underground Structures**	4
4.1	Investigation of Basic Geological Conditions	4
4.2	Investigation of Engineering Geological Properties	7
5	**Engineering Geological Investigation Methods for Underground Structures**	11
5.1	Engineering Geological Investigation Methods for Tunnels	11
5.2	Engineering Geological Investigation Methods for Underground Powerhouse System	13
6	**Engineering Geological Assessment on Surrounding Rock of Underground Structures**	17
6.1	Location Selection for Underground Structures	17
6.2	Surrounding Rock Classification and Engineering Geological Assessment of Surrounding Rock Stability	18
6.3	Prediction of Rock Burst	21
6.4	Prediction of Water Inrush and Mud Inrush	21
6.5	Prediction of Ground Temperature, Harmful Gases and Radioactivity	22
6.6	Prediction of External Water Pressure	24
6.7	Engineering Geological Assessment on Stability of Surrounding Rock for High Water Head Tunnels	25
6.8	Engineering Geological Assessment of Surrounding Rock Stability for Air Cushion Surge Chambers	26
7	**Engineering Geological Work During Underground Structure Construction**	28
7.1	Investigation for Special Engineering Geological Problems	28
7.2	Geological Work During Construction	28
Appendix A	Determination of Firmness Coefficient, Unit Elastic Resistance Coefficient and Strength-Stress Ratio of Surrounding Rock	31
Appendix B	Types of Soluble Halite	33
Appendix C	Empirical Values of Main Physical and Mechanical Parameters of Surrounding Rocks	34
Appendix D	Instability Mechanism Types and Failure	

	Modes of Surrounding Rock ··················	35
Appendix E	Allowable Relative Convergence Values for Tunnels and Caverns ····························	37
Appendix F	Maximum Allowable Concentration of Harmful Gases and Main Indexes of Air Composition for Underground Caverns ·····························	38
Appendix G	Classification of Gas Tunnel ·······················	39

Explanation of Wording in This Specification ···················· 40
List of Quoted Standards ·· 41

1 General Provisions

1.0.1 This specification is formulated with a view to standardizing the content, methods, assessment, and technical requirements for the engineering geological investigation of underground structures for hydropower projects, to ensure the investigation quality.

1.0.2 This specification is applicable to the engineering geological investigation of underground structures for large-sized hydropower projects. This specification is not applicable to the engineering geological investigation of underground structures for small- and medium-sized hydropower projects.

1.0.3 In addition to this specification, the engineering geological investigation of underground structures for hydropower projects shall comply with other current relevant standards of China.

2 Terms

2.0.1 surrounding rock

rock mass around an underground cavern, in which stresses are redistributed due to excavation; or rock mass that might affect the cavern stability and deformation

2.0.2 group of caverns

underground structure system consisting of multiple excavated caverns

2.0.3 hillside tunnel

tunnel on one side of a hill

2.0.4 watershed tunnel

tunnel through the mountain ridge

2.0.5 rock burst

dynamic phenomenon of explosive breaking and popping of rock walls caused by sudden release of stresses in the surrounding rock during cavern excavation under high in-situ stress

2.0.6 water inrush

groundwater gushing or spurting suddenly into the cavern during construction of underground cavern through an aquifer or permeable stratum

2.0.7 mud inrush

sudden gushing of mud in large volume into the cavern during construction of underground cavern through karst caves filled by argillaceous material or fault-fractured zones with a large mud content

2.0.8 high water head tunnel

tunnel with an internal water head no less than 100 m

2.0.9 deep tunnel

tunnel with a burial depth greater than 300 m

2.0.10 long tunnel

tunnel with a length greater than 2000 m

2.0.11 large-span underground cavern

underground cavern with a span greater than 20 m

3 Basic Requirements

3.0.1 The engineering geological investigation of underground structures shall ascertain the engineering geological conditions of underground structure area, analyze the engineering geological factors and problems affecting the stability of surrounding rock of underground caverns, assess the stability of surrounding rock of underground caverns, and provide engineering geological data for the design, excavation and surrounding rock support of underground structures.

3.0.2 The engineering geological investigation of underground structures shall proceed in stages: planning, pre-feasibility study, feasibility study, tender design, and detailed design. The work depth of investigation at each stage shall comply with the current national standard GB 50287, *Code for Hydropower Engineering Geological Investigation*.

3.0.3 The engineering geological investigation of underground structures at each stage shall be conducted according to the engineering geological investigation outline.

3.0.4 The engineering geological investigation of underground structures shall be conducted according to the investigation stage and the layout of structures, from ground to underground, from general investigation to special engineering geological problem study, and from qualitative assessment to quantitative assessment.

3.0.5 The raw data in the investigation shall be authentic, accurate and complete, and shall be timely sorted out and comprehensively analyzed.

3.0.6 The investigation shall meet the requirements for environmental protection.

4 Engineering Geological Investigation Content of Underground Structures

4.1 Investigation of Basic Geological Conditions

4.1.1 The topography and geomorphy, stratigraphy and lithology, geological structure, geophysical phenomena, karst, and hydrogeological conditions shall be ascertained in the investigation of basic geological conditions of the underground structure area.

4.1.2 The topographic and geomorphic investigation of the underground structure area shall meet the following requirements:

1. The investigation in the underground structure area shall ascertain the landform and genetic types, and analyze their relation with lithology, geological structure, and neotectonics.

2. The investigation along the tunnel shall ascertain the development degree and cutting depth of surface water system and gullies, and the variation in gully water flow.

3. The investigation in the underground powerhouse area shall ascertain the topographic and geomorphic features and the distribution, cutting depth and terrain integrity of gullies and valleys.

4.1.3 The stratigraphic and lithologic investigation of the underground structure area shall meet the following requirements:

1. For magmatic rock, the mineral composition, chemical composition, texture, primary structure, and lithofacies characteristics shall be ascertained, and the following requirements shall be met:

   1) For intrusive body and veins, the attitude, distribution scale, contact relation, and alteration characteristics of contact zone shall be ascertained.

   2) For extrusive rocks, the flow structure, zonation, eruption cycle, and contact relation with upper and lower strata shall be ascertained.

   3) The investigation of magmatic rocks shall focus on the alteration, eruption interruption, dyke, contact relation, etc.

2. For sedimentary rock, the mineral composition, chemical composition, texture and structure, cementation, variation in lithology and lithofacies, sediment rhythm, formation type, and stratigraphic contact relation shall be ascertained, focusing on the weak strata, soluble halite,

coal-bearing strata, swelling rock, and soluble halite distributed in the underground structure area.

3   For metamorphic rock, the mineral composition, chemical composition, texture, structure, metamorphic degree, and metamorphism type shall be ascertained, focusing on the investigation of weak strata including phyllite, slate, schist, etc.

4   The rock formation of the underground structure area shall be classified according to the genetic type, hardness, structural characteristics, stratified combination conditions, and physical and mechanical properties of rocks. The detailed degree of rock formation classification shall match the engineering geological mapping scale. The weak interlayer, swelling rock, soluble halite, and rock stratum with harmful gas and radioactive minerals may be shown on a larger scale.

5   The distribution, genetic type, thickness, stratification, and material composition of the Quaternary overburden shall be ascertained in the investigation of the portal section, hillside section with small margin and gully-crossing section of the underground structure.

4.1.4 The geological structural investigation of underground structure area shall meet the following requirements:

1   The investigation of geological structures shall ascertain the geotectonic location, and the distribution and scale of major folds and faults around.

2   The investigation of folds shall ascertain the attitude of strata and the morphological characteristics, scale and distribution of folds.

3   The investigation of faults shall ascertain the distribution, attitude, tectonites, fracture and influenced zone width of faults, group the faults by attitude, rank the faults by scale, and classify the faults by property. The discontinuity classification of rock mass shall comply with the current national standard GB 50287, *Code for Hydropower Engineering Geological Investigation*.

4   The investigation shall focus on the faults that have an major impact on the stability of surrounding rocks of the underground structures. When a tunnel passes through a potential active fault, the fault activity and its influence on tunnel works shall be studied. The study on fault activity shall comply with the current sector standard NB/T 35098, *Specification of Regional Tectonic Stability Investigation for Hydropower Projects*.

5   For joints and fissures, investigation and statistics should be conducted on the number of joint or fissure sets, dominant attitude, spacing, persistence, roughness and undulation, weathering and alteration of fissure surfaces, aperture, infill, groundwater conditions, and volumetric joint count of rock mass. The area of the statistic window shall not be less than 10 m^2, its position shall be typical geologically, and its orientation shall be considered.

4.1.5 The investigation of geophysical phenomenon of underground structure area shall meet the following requirements:

1   For weathering characteristics, ascertain the weathering degree and depth of rock mass in underground structure area, focusing on the distribution, thickness and characteristics of weathered zones in areas where the portal section of cavern, shallow-buried section and underground powerhouse locate. The classification of the weathering zones of rock mass shall comply with the current national standard GB 50287, *Code for Hydropower Engineering Geological Investigation*.

2   For the relaxation of rock mass, ascertain the relaxation degree and depth, focusing on the distribution, thickness and characteristics of relaxed zones in areas where the portal section, shallow-buried section, and underground powerhouse locate. The classification of the relaxed zones of rock mass shall comply with the current national standard GB 50287, *Code for Hydropower Engineering Geological Investigation*.

3   For the deformation and failure of slope, ascertain the distribution, scale, and development characteristics of collapse, landslide, deformation bodies, etc., focusing on the characteristics of slope deformation and failure in areas where the portal section, shallow-buried section, and underground powerhouse locate, and on the distribution and stability of the major collapse, landslide and deformation bodies along the tunnel.

4   For debris flow, ascertain the distribution, type, scale, basin characteristics, formation conditions, developing history, and developing trend, focusing on the development characteristics of debris flow near the portals.

5   For abandoned mines and goafs, ascertain their distribution, form and scale, and the characteristics of ground and underground deformation and failure.

6   For freeze-thaw weathering of rock and soil mass, ascertain the

distribution, scale, and characteristics of the freeze-thaw regolith and the crushed rock, freeze-thaw talus glacier, freeze-thaw debris flow, etc. generated by freeze-thaw weathering, focusing on the characteristics of the freeze-thaw weathering of rock and soil mass in the portal section and shallow-buried section.

4.1.6 The karst investigation shall be conducted in the underground structure area in accordance with the current sector standard NB/T 10075, *Specification for Karst Engineering Geological Investigation of Hydropower Projects*.

4.1.7 The investigation of hydrogeological conditions in the underground structure area shall meet the following requirements:

1 The investigation shall ascertain the basic type, water table, burial depth, pressure, volume, temperature, and hydrochemical composition of groundwater, and the water-bearing property and permeability of rock mass, and define the aquifer and aquiclude; analyze the recharge, runoff and discharge conditions of each aquifer considering the issues of spring, and define the hydrogeological units.

2 The investigation shall focus on the catchment conditions of the areas where the tunnel may pass through, including the syncline hinge, fault-fractured zone and their intersection, closely jointed zone, shallow-buried section, and gully-crossing section.

4.2 Investigation of Engineering Geological Properties

4.2.1 The investigation of physical and mechanical properties of surrounding rock shall meet the following requirements:

1 For physical and mechanical properties of rock, take sample and measure the density, water absorption, compressive strength, tensile strength, shear strength, point load strength, elastic modulus, Poisson's ratio, acoustic wave value, etc.

2 For mechanical properties of rock mass, measure on site the deformation modulus, shear strength, wave velocity value, etc. The unit elastic resistance coefficient of surrounding rock may be tested.

3 For mechanical properties of discontinuities, measure the shear strength of discontinuity, the deformation and seepage deformation parameters of weak interlayer, etc.

4 The firmness coefficient, unit elastic resistance coefficient and strength-stress ratio of surrounding rock should be determined in accordance with Appendix A of this specification.

4.2.2 The investigation of the initial in-situ stress in rock mass of underground structure shall meet the following requirements:

1. The magnitude and direction of the initial in-situ stress in rock mass shall be measured, and the zoning of initial in-situ stress in rock mass in the bank slope shall be conducted. The determination of the initial in-situ stress in rock mass shall comply with the current standards of China GB 50287, *Code for Hydropower Engineering Geological Investigation*; and DL/T 5367, *Code for Rock Mass Stress Measurements of Hydroelectric and Water Conservancy Engineering*.

2. The ranking of initial in-situ stress and the classification of rock mass deformation and failure and identification of rock burst under high in-situ stress shall be conducted. The classification and identification shall comply with the current national standard GB 50287, *Code for Hydropower Engineering Geological Investigation*.

4.2.3 The investigation of physical and mechanical properties of overburden for the underground structure shall measure the natural water content, density, deformation modulus, compression modulus, shear strength, permeability coefficient, etc.

4.2.4 The investigation of physical and mechanical properties of special rocks/soils shall meet the following requirements:

1. For soft rock, the genetic type shall be investigated, the natural water content, density, compressive strength, disintegration resistance index, free swelling ratio, etc. shall be measured, and the rheology properties may be tested and studied. The classification of the genetic type of soft rock shall comply with the current sector standard NB/T 10339, *Specification for Dam Site Engineering Geological Investigation of Hydropower Projects*.

2. For swelling rock, the mineral composition, chemical composition, cation exchange capacity, saturated water absorption, free swelling ratio, expansion rate under a certain pressure, expansive force, etc. shall be measured. The identification and classification of the geological characteristics of swelling rock shall comply with the current sector standard NB/T 10339, *Specification for Dam Site Engineering Geological Investigation of Hydropower Projects*.

3. For soluble halite, the solubility, dissolution collapsibility and salt expansion under the action of running water of soluble halite, and the corrosivity to concrete and metal structure shall be investigated.

The type of soluble halite shall comply with Appendix B of this specification.

4   For loess, the water content, liquid limit, plastic limit, shear strength, collapsibility coefficient, etc. shall be measured. The identification of loess collapsibility shall comply with the current sector standard NB/T 10339, *Specification for Dam Site Engineering Geological Investigation of Hydropower Projects*.

5   For frozen soil, the total water content, relative ice content, freezing temperature, heat conductivity coefficient, upper limit depth of perennially frozen soil, lower limit depth of seasonally frozen soil, etc. shall be measured, focusing on the frost-heave and thaw collapsibility of surrounding rock in frozen soil.

4.2.5 The investigation of ground temperature, harmful gas and radioactive source shall meet the following requirements:

1   The investigation of ground temperature shall collect the local ground temperature and geothermal data, and measure the temperature, heat conductivity coefficient, temperature conductivity coefficient, etc. of rock masses at different depths of the borehole and exploratory adit in the underground structure area.

2   The investigation of harmful gases shall collect the data on the distribution of strata that might produce and store gases, analyze the migration, accumulation and closure conditions of the harmful gases, and measure the composition and content of harmful gases.

3   The investigation of radioactive sources shall collect the regional geological data on the radioactive sources, and measure the equilibrium equivalent concentration and environmental radioactive radiation quantity of the radon and its progenies in the exploratory adit, borehole and construction adit.

4.2.6 The investigation of water inrush and mud inrush shall meet the following requirements:

1   The local precipitation data shall be collected, the distribution, surface water flow, etc. of gullies and karst depression along the tunnel shall be ascertained, focusing on the catchment area and conditions, and overburden and weathered broken rock distributions of the gullies and depressions.

2   The distribution and development characteristics of the syncline hinge,

weak rock zone, fault-fractured zone, closely jointed zone, etc., in the surrounding rock shall be ascertained.

3 The karst cave and corrosive closely jointed zone in the surrounding rock, and the filling features of gravelly soil shall be ascertained.

4 Dynamic observation shall be conducted on groundwater, spring and surface water.

4.2.7 The focus on investigation of surrounding rock of underground structure shall be determined according to the hardness and structural type of rock mass, and shall meet the following requirements:

1 For hard and intact rock mass, focus on testing and studying the in-situ stress state of rock mass and the strength-stress ratio of the rock, and analyzing the influence of high in-situ stress on the surrounding rock of caverns under excavation.

2 For fractured blocky hard rock mass, focus on ascertaining the development, combination form of all kinds of discontinuities, testing the physical and mechanical properties, and analyzing the influence of block combination on the local stability of surrounding rock.

3 For stratified hard rock mass, focus on investigating the bedding, interlayer squeezing dislocation zone, etc., testing the anisotropic characteristics of the mechanical properties, and analyzing their influence on surrounding rock stability.

4 For weak rock mass, focus on testing the clay mineral compositions, physical and mechanical properties, water-physical property, rheological property, etc., and analyzing their adverse effects on tunneling.

5 Engineering Geological Investigation Methods for Underground Structures

5.1 Engineering Geological Investigation Methods for Tunnels

5.1.1 The engineering geological mapping of the tunnel area shall meet the following requirements:

   1   The engineering geological mapping of the tunnel area shall be carried out based on available topographic and geological data and remote sensing interpretation. The method, scope and scale of the engineering geological mapping at each investigation and design stage shall comply with the current standards of China GB 50287, *Code for Hydropower Engineering Geological Investigation*; and NB/T 10074, *Specification for Engineering Geological Mapping of Hydropower Projects*.

   2   The engineering geological mapping results shall be checked based on the geological logging data of construction pilot tunnels and adits.

5.1.2 The exploration of tunnels shall meet the following requirements:

   1   The exploration shall be carried out at the tunnel portal, gully-crossing section, shallow-buried section and the sections where there might be potential major engineering geological problems.

   2   The adit, drilling and geophysical exploration should be adopted. Deep tunnels may adopt ultra-deep boreholes, and deep watershed tunnels may adopt ultra-long exploratory adits. Adit exploration shall comply with the current sector standard NB/T 10340, *Specification for Pit Exploration of Hydropower Projects*, the drilling exploration shall comply with the current sector standard NB/T 35115, *Specification for Drilling Exploration of Hydropower Projects*, and the geophysical exploration shall comply with the current sector standard NB/T 10227, *Code for Geophysical Exploration of Hydropower Projects*.

5.1.3 Tests of surrounding rock shall meet the following requirements:

   1   The surrounding rock of tunnel shall be sampled for physical and mechanical properties tests, thin section identification, and mineral and chemical composition analysis. The physical and mechanical properties tests of rock shall comply with the current sector standard DL/T 5368, *Code for Rock Tests of Hydroelectric and Water Conservancy Engineering*. The analysis of mineral and chemical composition of rock and soil masses shall comply with the current sector standard

NB/T 35102, *Specification for Soil Test in Situ in Borehole of Hydropower Projects*.

2   The tests of special surrounding rocks/soils shall meet the following requirements:

   1) Perform special tests of swelling properties and disintegration resistance for soft rock and swelling rock.

   2) Perform special tests of dissolution characteristics for soluble halite.

   3) Perform special tests for loess.

   4) Perform special tests for frozen soil.

3   If the tunnel passes through a potential active fault, the age of fault activity shall be determined in accordance with the current sector standard NB/T 35098, *Specification of Regional Tectonic Stability Investigation for Hydropower Projects*.

4   The rebound value test, elastic wave velocity test and geophysical logging of surrounding rock should be carried out by use of exploratory adits or pilot tunnel. The shear strength test, deformation test, elastic modulus test in hole, in-situ stress test and unit elastic resistance coefficient test of surrounding rock may be carried out. The mechanical properties test of rock mass shall comply with the current sector standard DL/T 5368, *Code for Rock Tests of Hydroelectric and Water Conservancy Engineering*. The test of in-situ stress in rock mass shall comply with the current sector standard DL/T 5367, *Code for Rock Mass Stress Measurements of Hydroelectric and Water Conservancy Engineering*. The geophysical prospecting test shall comply with the current sector standard NB/T 10227, *Code for Geophysical Exploration of Hydropower Projects*.

5   The tests of ground temperature, harmful gas and radioactivity of tunnel shall be carried out in the borehole, exploratory adit and pilot tunnel.

6   The groundwater and surface water shall be sampled for water quality analysis. The water quality analysis shall comply with the current sector standard NB/T 35052, *Specification of Water Quality Analysis for Hydropower Engineering Geological Investigation*.

7   The surrounding rock shall be subject to the water pressure test in borehole, and hydraulic fracturing and high-pressure water tests may be carried out. The water pressure tests in boreholes shall comply with the

current sector standard NB/T 35113, *Specification for Water Pressure Test in Borehole of Hydropower Projects*.

5.1.4 The dynamic long-term observation of groundwater in tunnel area shall be carried out, and the observation duration shall not be less than one hydrological year. The observation items should include groundwater table, water pressure, water volume, water temperature and water quality. Observation points should include boreholes, exploratory adits and springs.

5.1.5 The deformation observation and monitoring of surrounding rock should be carried out at the tunnel sections with complex geological conditions and important tunnel sections. The deformation observation and monitoring of surrounding rock shall comply with the current sector standards DL/T 5006, *Code for Rock Mass Observations of Hydroelectric and Water Conservancy Engineering*; and NB/T 35039, *Specification for Geological Observation of Hydropower Projects*.

5.2 Engineering Geological Investigation Methods for Underground Powerhouse System

5.2.1 The engineering geological mapping of the underground powerhouse system area shall meet the following requirements:

1 The engineering geological mapping of the underground powerhouse system area shall be carried out based on the available topographical and geological data. The method, scope and scale of the engineering geological mapping at each investigation and design stage shall comply with the current standards of China GB 50287, *Code for Hydropower Engineering Geological Investigation*; and NB/T 10074, *Specification for Engineering Geological Mapping of Hydropower Projects*.

2 The engineering geological mapping results of underground powerhouse system area shall be checked based on the geological logging data of construction pilot tunnels and adits.

5.2.2 The exploration of underground powerhouse system shall meet the following requirements:

1 The exploration of underground powerhouse system shall be conducted within a certain area where the surge shaft, high-pressure pipe, bifurcation, underground powerhouse cavern group and tailrace tunnel locate. The main exploration methods include adit, drilling and geophysical exploration. Adit exploration shall comply with the current sector standard NB/T 10340, *Specification for Pit Exploration of Hydropower Projects*. Drilling exploration shall comply with

the current sector standard NB/T 35115, *Specification for Drilling Exploration of Hydropower Projects*. Geophysical exploration shall comply with the current sector standard NB/T 10227, *Code for Geophysical Exploration of Hydropower Projects*.

2   Exploratory adits shall be arranged for the large-span underground cavern such as underground powerhouse. The exploratory adits should be excavated longitudinally and transversely near the abutment elevation of the proposed conventional underground powerhouse, and the adits should cross the proposed powerhouse cavern and extend to a distance of 1 times the sidewall height. For the underground powerhouse of a pumped storage power station, the exploratory adits should be excavated longitudinally and transversely at the position 30 m to 50 m above the proposed powerhouse cavern roof, and the adits shall extend to the position beyond the bifurcation with the highest water head and maximum burial depth.

3   Boreholes in different directions may be arranged in the exploratory adits for the large-span cavern of underground powerhouse depending on the complexity of geological conditions and the scale of the proposed underground cavern. The vertical boreholes shall extend to the position 10 m to 30 m below the design invert elevation of the cavern.

4   Exploratory adits and boreholes shall be arranged for both the conventional surge chamber and air cushion surge chamber. For high-pressure pipes, boreholes shall be arranged, and exploratory adits may be constructed.

5   Inter-adit or inter-borehole elastic wave or electromagnetic wave computed tomography (CT) should be conducted for the underground powerhouse cavern exploration.

5.2.3 Physical and mechanical tests of rock and rock mass of the underground powerhouse system shall meet the following requirements:

1   The surrounding rock in the underground powerhouse system shall be sampled for physical and mechanical properties tests, thin section identification, and mineral and chemical composition analysis. The shear strength test of discontinuity, triaxial strength test, rheological test, and special rock test may be performed. The physical and mechanical properties tests of rock shall comply with the current sector standard DL/T 5368, *Code for Rock Tests of Hydroelectric and*

Water Conservancy Engineering. The analysis of mineral and chemical composition of rock and soil masses shall comply with the current sector standard NB/T 35102, *Specification for Soil Test in Situ in Borehole of Hydropower Projects*.

2   Field tests of surrounding rock mass of underground powerhouse system shall be carried out in the exploratory adit. The test items should include the shear strength test and deformation modulus test of rock mass and discontinuity, and the acoustic wave velocity testing and seismic wave velocity testing of rock mass. The correlation between the wave velocity and the static deformation modulus of rock mass should be established. The rheological test of weak rock stratum may be carried out, and the unit elastic resistance coefficient of the surrounding rock of high-pressure pipe may be tested. The mechanical property test and wave velocity test of rock mass shall comply with the current sector standards DL/T 5368, *Code for Rock Tests of Hydroelectric and Water Conservancy Engineering* and NB/T 10227, *Code for Geophysical Exploration of Hydropower Projects*.

3   The initial in-situ stress of rock mass shall be tested in the exploratory adits of caverns, the stress relieving method and hydraulic fracturing technique may be adopted, and the regression analysis of in-situ stress field shall be conducted according to the test results. The in-situ stress test shall be carried out by the hydraulic fracturing technique in the exploratory adits of air cushion surge chamber and high-pressure pipes. Tests of in-situ stress in rock mass shall comply with the current sector standard DL/T 5367, *Code for Rock Mass Stress Measurements of Hydroelectric and Water Conservancy Engineering*.

4   At the detailed design stage, the surrounding rock of underground powerhouse system should be tested for the range and relaxation degree of relaxed zone under excavation and blasting by the elastic wave and borehole panoramic image methods.

5   The ground temperature, harmful gas content and radioactivity of the underground powerhouse system shall be tested in the borehole, exploratory adit and pilot tunnel.

5.2.4 Hydrogeological tests of underground powerhouse system shall meet the following requirements:

1   Water pressure tests in boreholes shall be carried out, and shall comply with the current sector standard NB/T 35113, *Specification for Water*

Pressure Test in Borehole of Hydropower Projects. For the area where the high-pressure pipe, high-pressure bifurcation, or air cushion surge chamber is arranged, the high-pressure water pressure test in borehole shall be carried out, the maximum test pressure shall not be less than 1.2 times the maximum design head or maximum design air pressure, and the hydraulic fracturing test may be carried out.

2   The groundwater table, pressure, volume and temperature in boreholes and exploratory adits shall be measured in the hydrogeological test.

3   The water quality analysis shall be conducted by groundwater and surface water sampling, and shall comply with the current sector standard NB/T 35052, *Specification of Water Quality Analysis for Hydropower Engineering Geological Investigation*.

5.2.5   The numerical simulation method may be used in the groundwater seepage field study of the underground powerhouse system.

5.2.6   The dynamic observation of groundwater in the underground powerhouse system area shall be carried out, and the observation duration shall not be less than one hydrological year. The observation items shall include groundwater table, pressure, volume, temperature and quality. Observation points shall include boreholes, exploratory adits and springs.

5.2.7   The deformation observation and monitoring of surrounding rock should be carried out for the underground powerhouse system. The deformation observation should be carried out in exploratory adits and test tunnels, and the deformation monitoring design shall be carried out at the detailed design stage. The deformation observation and monitoring of surrounding rock shall comply with the current sector standards DL/T 5006, *Code for Rock Mass Observations of Hydroelectric and Water Conservancy Engineering*; and NB/T 35039, *Specification for Geological Observation of Hydropower Projects*.

6 Engineering Geological Assessment on Surrounding Rock of Underground Structures

6.1 Location Selection for Underground Structures

6.1.1 Tunnel alignment selection shall meet the following requirements:

1. Tunnel alignment shall be selected according to the engineering geological conditions of the tunnel area, hydraulic tunnel design, and general layout of project. The areas with complex engineering geological and hydrogeological conditions unfavorable to the surrounding rock stability of the tunnel should be avoided, and a shorter route should be selected. For alignment selection of long tunnels, the tunneling conditions of adits and the stability conditions of tunnel portals shall be considered.

2. The tunnel alignment should avoid traversing the negative landforms such as gully, saddle on a mountain ridge and large karst depression. When a tunnel crosses gullies or a hillside tunnel is shallowly buried, geological recommendations should be provided for tunnel alignment selection according to the overburden thickness, material composition and slope stability conditions of the gully section or the hillside with small margin.

3. The tunnel alignment should avoid large fault-fractured zones, active faults, soluble halite, swelling rock, development zone of karst cave, goaf, enrichment region of harmful gas and radioactive mineral source, and groundwater confluence areas such as water-conducting fault and underground karst river, and shall meet the following requirements:

    1) The included angle between the tunnel alignment and the strike of the main fault or special rock zone should be greater than 30°.

    2) When the tunnel crosses an active fault, the age, mode and rate of fault activity shall be studied to predict the maximum possible accumulative creep displacement or the maximum possible sudden displacement in the design service life, and provide the geological data for tunnel design.

4. Tunnel portals should be arranged on the area with intact terrain, stable slope, exposed bedrock, hard rock, and weak weathering and stress relief, or the area with shallow overburden.

6.1.2 The site selection of underground powerhouse system shall meet the following requirements:

1   The underground powerhouse system should be arranged in the stable mountain with intact terrain.

2   The powerhouse site should be located at the region with hard and intact rock mass and simple hydrogeological conditions. The underground powerhouse should be arranged in the normal in-situ stress zone and shall avoid regional faults, active faults, goaf, strongly weathered and relaxed rock mass, large karst caves, underground rivers, etc.

3   The entrance of the underground powerhouse system should avoid the area with the distribution and impact of adverse geological-physical phenomena such as landslide, collapse, deformed rock mass and debris flow.

6.1.3 Based on the powerhouse site selection, the underground powerhouse axis orientation shall be determined according to the development characteristics of discontinuities, the state of in-situ stress and the layout of the main structures in the plant area. The following requirements should be met:

1   The included angle between the powerhouse axis and the strike of the stratum, fault or main fissure set should not be less than 60°.

2   In high in-situ stress areas, the included angle between the powerhouse axis and the maximum principal stress direction of the initial in-situ stress in rock mass in the plant area should not be greater than 30°.

6.2 Surrounding Rock Classification and Engineering Geological Assessment of Surrounding Rock Stability

6.2.1 The surrounding rock classification for underground structures shall include preliminary classification and detailed classification, and shall assess the overall stability of surrounding rock and give suggestions for support system according to the classification results. The surrounding rock classification shall comply with the current national standard GB 50287, *Code for Hydropower Engineering Geological Investigation*.

6.2.2 The preliminary classification shall be conducted according to the factors such as the rock hardness, rock mass structure type, and rock mass integrity.

6.2.3 The detailed classification shall, based on the preliminary classification and engineering geological sectioning of surrounding rock, be conducted by taking the sum of the ratings of 5 factors including rock strength, rock mass integrity, discontinuity condition, groundwater condition and attitude of the main discontinuity as the basic criterion, and the strength-stress ratio of surrounding rock as the limit criterion.

6.2.4 The determination of physical and mechanical parameters of surrounding rock mass and discontinuities shall comply with the current national standard GB 50287, *Code for Hydropower Engineering Geological Investigation*, and shall meet the following requirements:

1 The recommended values for geological parameters of the physical and mechanical properties of surrounding rock shall be proposed by performing the deformation and shear strength tests of rock mass and discontinuities and analyzing their results taking into account the engineering geological classification of the surrounding rock and engineering geological analogy.

2 The permeability of the surrounding rock shall be proposed based on the results of water pressure tests in borehole. The permeability and seepage deformation parameters of the weak zones in the surrounding rock shall be proposed based on the results of high-pressure water pressure tests in borehole and hydraulic fracturing tests.

3 The magnitude and direction of initial in-situ stress in rock mass shall be determined comprehensively according to the initial in-situ stress test of rock mass, regression analysis for initial in-situ stress field, geological analysis for high in-situ stress phenomenon, regional tectonic stress analysis, topographic and geomorphic analysis, etc.

4 The anisotropy parameters of bedded surrounding rock deformation should be determined based on the test results of deformation perpendicular to and parallel to the bedding, the loading features of the surrounding rock, etc.

5 The empirical values of the main physical and mechanical parameters of surrounding rock should be determined in accordance with Appendix C of this specification.

6.2.5 The engineering geological analysis of the surrounding rock stability shall meet the following requirements:

1 The analysis of geological factors affecting surrounding rock stability shall be conducted from four aspects: rock hardness, rock mass structure, groundwater, and in-situ stress state of rock mass.

2 The overall stability of surrounding rock shall be assessed based on the engineering geological classification of the surrounding rock. The surrounding rock classification for the large-span underground

powerhouse cavern shall be conducted by parts, and the surrounding rock stability shall be assessed respectively based on the classification in different parts.

3   The local stability of surrounding rock should be assessed by block analysis.

4   The surrounding rock instability failure mechanisms should be studied by the investigation of collapses in exploratory adits, construction pilot tunnels, adits and the caverns under excavation, to analyze the control factors of failure, and mechanical mechanisms and failure modes of the deformation and failure of surrounding rock. The instability mechanism types and failure modes of surrounding rock should be in accordance with Appendix D of this specification.

5   The engineering geological analysis of the surrounding rock stability shall focus on the impact of the moderately and steeply dipping discontinuities on the surrounding rock stability at the sidewall and end wall, and the impact of gently dipping discontinuities on the surrounding rock stability at the crown.

6.2.6 In the calculation and analysis of the local stability of surrounding rock, when there exist unfavorable combined blocks of weak discontinuities in surrounding rock with low stress, the block limit equilibrium analysis and key block stability estimation may be conducted considering only the effect of gravity. For surrounding rock with loose medium, Protodyakonov's ground pressure theory may be used to calculate the height of possible collapse arch and the ground pressure.

6.2.7 Field tests and analysis of surrounding rock deformation should meet the following requirements:

1   In-situ monitoring should be conducted for the assessment of relaxation characteristics and stability of surrounding rock in excavation period, to monitor the elastic wave velocity, deformation, secondary stress of surrounding rock and the stress and strain of support structure.

2   The allowable relative convergence value of cavern periphery may be determined according to the site monitoring results of surrounding rock deformation, and may be taken as the criterion of surrounding rock stability. The allowable relative convergence values for tunnels or caverns should be in accordance with Appendix E of this specification.

6.2.8 The stability assessment of caverns in special rock and soil should study the rheological characteristics of soft rock, the swelling characteristics of swelling rock, the corrosion and dissolution characteristics of soluble halite, the collapsible characteristics of loess, the freeze-thaw characteristics of frozen soil, as well as the compactness, deformation and strength properties, and permeability characteristics of overburden, to provide basis for special treatment.

6.3 Prediction of Rock Burst

6.3.1 The geological factors triggering rock burst shall be analyzed from the aspects of lithology and rock mass strength, structural characteristics and integrity of rock mass, in-situ stress magnitude and direction, groundwater activity, etc.

6.3.2 Early prediction of rock burst shall, based on the engineering geological sectioning and classification of surrounding rock, be carried out according to the magnitude of the maximum principal stress in rock mass, strength-stress ratio of rock, high in-situ stress relief during excavation of exploratory adit, formation of disk-shaped rock cores, etc. The initial in-situ stress in rock mass shall be classified according to the magnitude of the maximum principal stress and the strength-stress ratio of rock. The classification of initial in-situ stress in rock mass, and the classification and identification of deformation and failure of rock mass under high in-situ stress shall be in accordance with the current national standard GB 50287, *Code for Hydropower Engineering Geological Investigation*.

6.3.3 Prediction of rock burst during excavation should, based on the check of engineering geological sectioning and classification of surrounding rock, be checked through comprehensive analysis according to the test and monitoring data on rock burst, groundwater activity, secondary stress of surrounding rock, microseism, acoustic emission characteristic, radon release, etc.

6.3.4 The rock burst intensity shall be classified according to the main phenomena, rock strength-stress ratio and the critical burial depth of rock burst. The intensity classification of rock burst shall be in accordance with the current national standard GB 50287, *Code for Hydropower Engineering Geological Investigation*.

6.3.5 Risk assessment of rock burst may be conducted on the basis of rock burst prediction. The risk assessment of rock burst shall comply with the current sector standard NB/T 10143, *Technical Code for Rockburst Risk Assessment of Hydropower Projects*.

6.4 Prediction of Water Inrush and Mud Inrush

6.4.1 The prediction of water inrush shall meet the following requirements:

1 The water inrush prediction shall, based on the hydrogeological investigation of the underground structure area, analyze the burial depth of tunnel, groundwater table, aquifer and permeable zones such as tunnel syncline water catchment structures, fault-fractured zones and its influenced zones, closely jointed zones, karst channels and their relation with surface gullies.

2 The water inrush prediction shall observe the surface gully water and spring water flow, analyze the groundwater recharge, runoff and discharge condition, and define the hydrogeological units.

3 The water inrush prediction should estimate the maximum inflow and stable inflow at the possible inrush sections such as sections crossing gullies, sections in fault-fractured zone and sections with karst development by the hydrogeological analogy method, water balance method, groundwater dynamics method or three-dimensional seepage field analysis based on the hydrological and meteorological data.

4 The water inrush prediction shall assess the harm of water inrush to the project, and propose the treatment suggestions.

6.4.2 The prediction of mud inrush shall meet the following requirements:

1 The mud inrush prediction shall, based on the investigation of surface gullies, karst depression, ponor and their filling thickness and features, surface collapse characteristics, as well as the thickness and properties of fault-fractured zone and weak rock zone, analyze the distribution and filling features of karst cave pipe system, fault-fractured zone and weak rock zone in underground structure area, and study the connectivity of the karst cave pipe, fault-fractured zone, and weak rock zone where the underground structures cross with the surface depression, ponor, and gully water.

2 The mud inrush prediction should further predict the tunnel sections where mud inrush might occur and the likely scales based on the result of water inrush prediction.

3 The mud inrush prediction shall assess the harm of mud inrush to the project, and propose treatment suggestions.

6.5 Prediction of Ground Temperature, Harmful Gases and Radioactivity

6.5.1 The prediction of ground temperature shall meet the following requirements:

1   The pertinent data on local ground temperature and geotherm, local ground temperature gradient and ground average annual temperature may be collected, and the ground temperature of the tunnel sections at different burial depths may be estimated.

2   The ground temperature gradient value should be determined based on the ground temperature data measured in boreholes and exploratory adits at different depths, to predict the ground temperature of the cavern.

3   In the deep long tunnel area with complex stratigraphy and lithology, topography, and geological structure conditions, the ground temperature prediction may, according to the heat conductivity coefficient, temperature conductivity coefficient of rocks and faults and hydrogeological conditions, simulate the geothermal field of the tunnel area using the numerical analysis method, analyze the characteristics of the geothermal field, and predict the ground temperature of the cavern.

4   The heat harm of ground temperature may, based on the ground temperature prediction, be classified according to the temperature adaptation level of personnel and equipment, and the suggestions on ventilation and cooling shall be given.

6.5.2   The prediction of harmful gases shall meet the following requirements:

1   The harmful gas prediction shall, based on the local geological data, analyze the geological environment generating harmful gases, evaluate and predict the harm degree according to the harmful gas content testing results in accordance with the relevant national and sector standards, and propose the harm mitigation solutions. The maximum allowable concentration of harmful gases and main indexes of air composition for underground caverns shall be in accordance with Appendix F of this specification.

2   When underground structures cross coal-, oil-, and gas-bearing strata, or there are similar strata in the surrounding area, harmful gas prediction shall, based on the analysis of the generating environment and migration and accumulation conditions of gas, assess and predict the harm degree of gas according to the relevant national and sector standards and the testing results of gas content, pressure, and emission rate, and propose the harm mitigation solutions. The classification of gas tunnel shall comply with Appendix G of this specification.

6.5.3   The radioactivity prediction shall, based on the investigation of stratigraphy, lithology, and geological structures and the radioactivity test results, analyze the potential radiation hazard and propose the mitigation solutions. The radioactivity control standard of underground caverns shall comply with the current standards of China GB 18871, *Basic Standards for Protection Against Ionizing Radiation and for the Safety of Radiation Sources*; and GBZ 116, *Standard for Controlling Radon and Its Progenies in Underground Space*.

6.6 Prediction of External Water Pressure

6.6.1 The prediction of external water pressure shall meet the following requirements:

1. The stratigraphy and lithology, geological structure, rock mass permeability, groundwater activity, groundwater level, and the recharge, runoff, and discharge conditions in the cavern and its surrounding area shall be analyzed.

2. The groundwater level should be directly observed via boreholes, springs, wells, gullies, etc. and the observation data of the highest groundwater level should be obtained. When no measured data is available, the groundwater level may be determined by comprehensive analysis according to terrain relief, rock mass weathering and relaxation characteristics, etc.

3. The external water pressure prediction in the karst area shall also study the karst development characteristics and the hydraulic relation between surface water and groundwater in the cavern and its surrounding area.

6.6.2 The determination of external water pressure shall meet the following requirements:

1. The external water pressure should be estimated by groundwater head reduction according to the groundwater level above the cavern, overlying rock mass permeability, and groundwater activity in the overlying rock mass. The empirical value of the external water pressure reduction factor shall be in accordance with the current national standard GB 50287, *Code for Hydropower Engineering Geological Investigation*.

2. When there are multiple hydrogeological structures, groundwater in the surrounding rock has a weak relation to the upper layer groundwater, and the groundwater level in the surrounding rock of a concrete-lined cavern is low, the external water pressure value may be determined by the internal water head.

6.7 Engineering Geological Assessment on Stability of Surrounding Rock for High Water Head Tunnels

6.7.1 The engineering geological assessment on surrounding rock stability for a steel-lined high water head tunnel should meet the following requirements:

1 When the steel lining and surrounding rock share the internal water pressure, the surrounding rock capacity against radial deformation should be assessed using the unit elastic resistance coefficient.

2 The unit elastic resistance coefficient may be calculated using the elastic or deformation modulus and Poisson's ratio of the rock mass in accordance with Appendix A of this specification or may be determined by the surrounding rock classification analogy. The empirical value of the unit elastic resistance coefficient of different surrounding rocks should be in accordance with Appendix C of this specification, it may also be determined by in-situ tests.

6.7.2 The engineering geological assessment on surrounding rock stability of a concrete-lined high water head tunnel shall meet the following requirements:

1 The stability assessment of overlying rock mass against lifting shall be conducted for a concrete-lined high water head tunnel. The minimum thickness of the overlying rock mass on a pressure tunnel shall be such that the tunnel hydrostatic pressure is less than the rock mass weight above the tunnel roof, and can be calculated by the following formula:

$$C_{RM} = \frac{h_s \gamma_w F}{\gamma_r \cos \alpha} \qquad (6.7.2)$$

where

C_{RM} is the minimum thickness of the overlying rock mass (m), excluding the thickness of completely weathered zone, highly weathered zone, and strongly relaxed zone;

h_s is the hydrostatic water head in the tunnel (m);

γ_w, γ_r are the unit weights of the water and the rock, respectively (kN/m³);

α is the river bank slope angle (°), when $\alpha > 60°$, $\alpha = 60°$;

F is the empirical coefficient, taken as 1.3 to 1.5.

2 The assessment of surrounding rock stability against hydraulic fracturing shall be carried out for a concrete-lined high water head tunnel. The hydrostatic pressure in the pressure tunnel shall be less than

the minimum principal stress of the surrounding rock. The minimum principal stress should be determined by the hydraulic fracturing technique.

3   The assessment of surrounding rock stability against leakage shall be carried out for a concrete-lined high water head tunnel. The surrounding rock should be Class Ⅰ or Class Ⅱ impermeable or slightly permeable rock mass, or the permeability of the surrounding rock mass shall be less than 1.0 Lu after high-pressure grouting.

4   When the concrete lining and surrounding rock share the internal water pressure, the surrounding rock capacity against radial deformation may be assessed using the unit elastic resistance coefficient. The value of the unit elastic resistance coefficient may be in accordance with Article 6.7.1 of this specification.

6.8   Engineering Geological Assessment of Surrounding Rock Stability for Air Cushion Surge Chambers

6.8.1   The content and methods of engineering geological assessment for the surrounding rock stability of the air cushion surge chamber with bolt-shotcrete support or concrete lining shall be in accordance with Article 6.7.2 of this specification, and shall meet the following requirements:

1   The surrounding rock of air cushion surge chamber should be medium-hard or hard, which should not be inferior to Class Ⅲ.

2   The weight of the overlying rock mass with a thickness of the minimum burial depth after deducting the slope overburden, completely weathered rock mass, highly weathered rock mass, and highly relaxed rock mass shall be greater than or equal to 1.3 to 1.5 times the design hydrostatic water pressure in the air cushion surge chamber and should be greater than 1.1 times the design air pressure. When there is a water curtain seal, the weight shall also be greater than the design water curtain pressure.

3   The minimum principal stress, σ_3, of the rock mass of the air cushion surge chamber shall be greater than or equal to 1.2 to 1.5 times the design air pressure in the air cushion surge chamber. When there is a water curtain seal, the minimum principal stress shall also be greater than the design water curtain pressure.

4   When the air cushion surge chamber is sealed by water curtain seal or surrounding rock obturator, the permeability of the rock mass should be

less than 1 Lu after grouting.

6.8.2 The ladle sealing structure and steel cage sealing structure may be applied when the engineering geological conditions of the air cushion surge chamber fail to satisfy Article 6.8.1 of this specification. The permeability of the rock mass with a cage-sealing structure should be less than 5 Lu.

6.8.3 The surrounding rock stability assessment of the air cushion surge chamber shall be carried out based on the engineering geological conditions, classification, in-situ stress, and high-pressure water and air permeability properties of the surrounding rock in the area where the air chamber and water curtain gallery are located, which are revealed during investigation and early excavation. The basis for determining the location, axis, type, and surrounding rock treatment measures of the air cushion surge chamber shall be given.

7 Engineering Geological Work During Underground Structure Construction

7.1 Investigation for Special Engineering Geological Problems

7.1.1 If there are large deformations of surrounding rock or other major engineering geological problems in some tunnel sections or parts of an underground structure, special investigation for engineering geological problems shall be carried out.

7.1.2 The special investigation for engineering geological problems shall check the deformation or boundary conditions for potential instability and relevant parameters, analyze the mechanism and type of deformation and failure, ascertain the engineering geological problems, and propose the treatment suggestions.

7.1.3 The method and workload of special investigation for engineering geological problems shall be determined according to the complexity of geological problems, field conditions, etc., and the investigation method shall meet the following requirements:

1 The geological data collection and logging shall be performed using various excavation faces.

2 Engineering geological mapping shall be adopted, and exploration and test should be performed.

3 Comprehensive geological analysis shall be carried out based on monitoring and testing data.

7.1.4 The investigation report on special engineering geological problems for the underground structure shall be submitted upon completion of the special investigation.

7.2 Geological Work During Construction

7.2.1 The geological work during construction of an underground structure may be carried out in two periods, the excavation period and the period after completion of the final section.

7.2.2 The geological work during the excavation period shall meet the following requirements:

1 The data on geological phenomena revealed during excavation shall be collected and logged.

2 The geological patrol inspection shall be carried out along with

excavation.

3   The surrounding rock tests may be carried out.

4   The geological prediction and forecast shall be carried out.

5   The engineering geological classification and sectioning of surrounding rock shall be checked.

6   The geological work in the excavation period shall include participation in the surrounding rock support design.

7.2.3 The geological work in the period after completion of the final section shall meet the following requirements:

1   The engineering geological mapping of surrounding rock shall be carried out.

2   The engineering geological classification and physical and mechanical property parameters of surrounding rock shall be checked.

3   The engineering geological description of surrounding rock shall be prepared.

4   The quality assessment of surrounding rock shall be carried out. The geological work shall include participation in the acceptance of surrounding rock and the provision of the geological comments.

7.2.4 The geological logging shall be carried out along with excavation and expanding excavation, and the geological data shall be collected by geological patrol inspection, observation, sketch, measurement, photography, etc., to provide the basis for the assessment on surrounding rock relaxation, deformation and stability, and the optimization of supporting scheme.

7.2.5 The check tests during the construction period include lab test on the physical and mechanical properties of rock, water quality analysis, and the deformation modulus test, shear strength test and in-situ stress test of surrounding rock.

7.2.6 The observation and monitoring during the construction period include groundwater inrush, surrounding rock deformation and failure, high in-situ stress, rock burst, ground temperature, harmful gases, and radioactivity.

7.2.7 The geological logging and mapping, sampling and test, observation and forecast, as well as assessment and acceptance of underground structures shall comply with the current sector standard NB/T 35007, *Geological Code of Construction Period for Hydropower Project*.

7.2.8 The geological data collected and logged during the construction period shall be sorted out and archived. The archived data should include the following:

1. The outline, technical specification and requirements of the geological work.

2. Original records such as mapping and logging, base map and geological description of the geological work.

3. Various survey data, test results, monitoring and testing data, special study results of geological work.

4. Photos and videos of geological work.

5. Construction geological briefing, construction geological forecast, and attached drawings.

6. Rock/soil samples and other materials requiring long-term preservation.

7. Geological monitoring requirements in the operation period.

8. Construction geological log, meeting minutes and decisions on major geological problems.

9. Approval documents from competent authorities, meeting minutes, relevant appraisal conclusions and consultation opinions on the engineering geological problems of surrounding rock.

10. Technical correspondences with various parties such as the designer, construction contractor, supervisor, and owner.

11. Geological data on design alteration and optimization.

12. Engineering geological investigation reports and special reports at all stages, investigation report of specialized engineering geological problems in the construction period, geological description for single item works acceptance, geological self-check report in safety appraisal and project acceptance, geological completion report and attached drawings of underground structures area.

13. Archive directory and instruction.

Appendix A Determination of Firmness Coefficient, Unit Elastic Resistance Coefficient and Strength-Stress Ratio of Surrounding Rock

A.0.1 The firmness coefficient of surrounding rock may be determined as follows:

1. The firmness coefficient of surrounding rock can be estimated by the following formula:

$$f = \alpha \frac{R_b}{10} \tag{A.0.1}$$

where

f is the firmness coefficient of surrounding rock;

R_b is the saturated uniaxial compressive strength of rock (MPa);

α is the correction factor, which is less than or equal to 1, depending on the strength and intactness of surrounding rock.

2. The firmness coefficient of surrounding rock may be determined according to the classification of the surrounding rock or engineering analogy.

A.0.2 The unit elastic resistance coefficient of surrounding rock may be determined as follows:

1. The unit elastic resistance coefficient of surrounding rock can be estimated by the following formula:

$$K_0 = \frac{E}{(1+\mu)\alpha} \tag{A.0.2}$$

where

K_0 is the unit elastic resistance coefficient of surrounding rock (MPa/cm);

E is the elastic modulus or deformation modulus of surrounding rock (MPa);

μ is the Poisson's ratio of surrounding rock;

α is the constant value, taken as 100 cm.

2. In the special test hole, the unit elastic resistance coefficient of

surrounding rock may be tested by radial hydraulic pressure pillow method or water pressure method.

3　The unit elastic resistance coefficient of surrounding rock may be determined according to the classification of surrounding rock and existing project analogy.

A.0.3　The strength-stress ratio of surrounding rock can be calculated by the following formula:

$$S = \frac{R_b K_v}{\sigma_m} \tag{A.0.3}$$

where

S　　is the strength-stress ratio of surrounding rock;

R_b　is the saturated uniaxial compressive strength of rock (MPa);

K_v　is the intactness coefficient of rock mass, i.e. the square of the P-wave velocity ratio between rock mass and rock;

σ_m　is the maximum principal stress of surrounding rock (MPa). In the absence of measured data, it may be determined in accordance with the current national standard GB 50287, *Code for Hydropower Engineering Geological Investigation*.

Appendix B Types of Soluble Halite

Table B Types of soluble halite

Type	Representative mineral	Basic property
Sulfates	Gypsum, anhydrite, mirabilite	Easy to dissolve, and anhydrite swelling with water
Chlorine salts	Rock salt, sylvinite	High solubility, obvious hygroscopicity, no change in volume during recrystallization
Carbonates	Soda	Great impact on disintegration rate, with alkaline taste
Potassium magnesium salts	Epsom salt, hexahydrite	High solubility

Appendix C Empirical Values of Main Physical and Mechanical Parameters of Surrounding Rocks

Table C Empirical values of main physical and mechanical parameters of surrounding rocks

Class of surrounding rock	Density ρ (t/m^3)	Friction coefficient f'	Cohesion c' (MPa)	Deformation modulus E_0 (GPa)	Possion's ratio μ	Firmness coefficient f	Unit elastic resistance coefficient K_0 (MPa/cm)
I	$\rho \geq 2.7$	$1.3 \leq f' < 1.5$	$1.8 \leq c' < 2.2$	$E_0 \geq 20$	$0.17 < \mu \leq 0.22$	$f \geq 7$	$K_0 \geq 70$
II	$2.5 \leq \rho < 2.7$	$1.1 \leq f' < 1.3$	$1.3 \leq c' < 1.8$	$10 \leq E_0 < 20$	$0.22 < \mu \leq 0.25$	$5 \leq f < 7$	$50 \leq K_0 < 70$
III	$2.3 \leq \rho < 2.5$	$0.7 \leq f' < 1.1$	$0.6 \leq c' < 1.3$	$5 \leq E_0 < 10$	$0.25 < \mu \leq 0.30$	$3 \leq f < 5$	$30 \leq K_0 < 50$
IV	$2.1 \leq \rho < 2.3$	$0.5 \leq f' < 0.7$	$0.3 \leq c' < 0.6$	$1 \leq E_0 < 5$	$0.30 < \mu \leq 0.35$	$1 \leq f < 3$	$5 \leq K_0 < 30$
V	$\rho < 2.1$	$0.35 \leq f' < 0.5$	$c' < 0.3$	$E_0 < 1$	$\mu > 0.35$	$f < 1$	$K_0 < 5$

Appendix D Instability Mechanism Types and Failure Modes of Surrounding Rock

Table D Instability mechanism types and failure modes of surrounding rock

Instability mechanism type	Failure mode		Mechanical mechanism	Rock hardness	Structure of rock mass
Strength-stress control	Brittle fracture	Rock burst	Sudden brittle failure caused by high concentration of compressive stress	Hard rock	Blocky and thick-bedded
		Splitting and spalling	Crack caused by compressive stress concentration		
		Tension crack collapse	Tensile failure caused by tensile stress concentration		
	Bending break		Bending crack caused by compressive stress concentration	Hard rock	Stratified and thin-bedded
	Plastic extrusion		Extruding into tunnel due to its stress exceeding the yield strength	Weak interlayer	Interbedded
	Internal extrusion collapse		Decrease of surrounding rock strength due to its confining pressure releasing and swelling by water absorption	Swelling soft rock	Stratified
	Loosening collapse		Loosening collapse under the action of gravity and tensile stress	Soft rock hard rock	Loose, cataclastic and block-fractured

Table D *(continued)*

Instability mechanism type	Failure mode	Mechanical mechanism	Rock hardness	Structure of rock mass
Weak discontinuity control	Block sliding and collapse	Instability of block under gravity	Hard rock (combination of weak discontinuities)	Blocky and stratified
Multi-factor control	Breaking and loosening	Shear breaking and loosening caused by compressive stress concentration	Hard rock (dense discontinuities)	Cataclastic, block-fractured and mosaic
	Shear slip	Slip and tensile crack caused by compressive stress concentration	Hard rock (combination of discontinuities)	Blocky and stratified

Appendix E Allowable Relative Convergence Values for Tunnels and Caverns

Table E Allowable relative convergence values for tunnels and caverns Δ (%)

Class of surrounding rock	Burial depth of tunnel H (m)		
	$H \leq 50$	$50 < H \leq 300$	$300 < H \leq 500$
III	$0.10 \leq \Delta < 0.30$	$0.20 \leq \Delta < 0.50$	$0.40 \leq \Delta < 1.20$
IV	$0.15 \leq \Delta < 0.50$	$0.40 \leq \Delta < 1.20$	$0.80 \leq \Delta < 2.00$
V	$0.20 \leq \Delta < 0.80$	$0.60 \leq \Delta < 1.60$	$1.00 \leq \Delta < 3.00$

NOTES:

1. The relative convergence value refers to the ratio of the measured displacement value to the distance between two testing points, or the ratio of the measured displacement of crown to the tunnel span.
2. For brittle surrounding rock, the smaller value is taken; and for plastic surrounding rock, the larger value is taken.
3. This table is applicable to the tunnels and caverns meeting the following conditions: the depth-span ratio is greater than or equal to 0.8 and less than or equal to 1.2; the burial depth is less than 500 m; the span is not greater than 20 m for Class III surrounding rock, not greater than 15 m for Class IV surrounding rock, and not greater than 10 m for Class V surrounding rock. Otherwise, the allowable relative convergence value shall be corrected according to project analogy.

Appendix F Maximum Allowable Concentration of Harmful Gases and Main Indexes of Air Composition for Underground Caverns

F.0.1 The maximum allowable concentration of harmful gases for underground caverns shall be in accordance with Table F.0.1.

Table F.0.1 Maximum allowable concentration of harmful gases for underground caverns

Name of gas	Symbol	Maximum allowable concentration	
		Volume ratio (%)	Mass-volume concentration (mg/m^3)
Carbon monoxide	CO	0.00240	30
Oxynitride	Converted to NO_2	0.00025	5
Sulfur dioxide	SO_2	0.00050	15
Ammonia	NH_3	0.00400	30
Hydrothion	H_2S	0.00066	10

F.0.2 The volume ratio of oxygen in the air near the working face of underground caverns shall not be lower than 20 %, the volume ratio of carbon dioxide shall not be higher than 0.5 %.

Appendix G Classification of Gas Tunnel

G.0.1 The gas tunnel may be classified into micro-content gas tunnel, low-content gas tunnel, high-content gas tunnel and gas outburst tunnel. The class of gas tunnel shall be determined by the highest grade of the working area.

G.0.2 The gas tunnel working area may be classified into non-gas working area, micro-content gas working area, low-content gas working area, high-content gas working area and gas outburst working area.

G.0.3 The micro-content gas working area, low-content gas working area and high-content gas working area may be identified by the absolute amount of gas effusion. The working area may be defined as the micro-content gas working area when the gas emission amount of the whole working area is greater than 0.0 m^3/min and less than 0.5 m^3/min, as the low-content gas working area when it is greater than 0.5 m^3/min and less than 1.5 m^3/min, and as the high-content gas working area when it is no less than 1.5 m^3/min.

G.0.4 Where there is a gas outburst risk in any place of the tunnel, the place shall be defined as the gas outburst working area, and the indicators of gas outburst shall be in accordance with Table G.0.4 simultaneously.

Table G.0.4 Indicators of gas outburst

Gas pressure P (MPa)	Initial velocity of gas diffusion ΔP (mmHg)	Firmness coefficient of coal f	Destructive type of coal
≥ 0.74	≥ 10	≤ 0.5	III, IV, V

NOTE Type III refers to cataclastic coal, Type IV refers to granulated coal, and Type V refers to mylonitized coal.

G.0.5 The grade of gas working area shall be defined by the gas content or gas pressure. The grade of gas working area shall be determined as per Table G.0.5.

Table G.0.5 Grade of gas working area

Grade	Gas content W_0 (m^3/t)	Gas pressure P (MPa)
Non-gas	$W_0 = 0$	$P = 0$
Micro-content gas	$0 < W_0 < 0.5$	$0 < P < 0.10$
Low-content gas	$0.5 \leq W_0 < 1.0$	$0 < P < 0.10$
High-content gas	$1.0 \leq W_0 < 8.0$	$0.10 \leq P < 0.74$
Gas outburst	$W_0 \geq 8.0$	$P \geq 0.74$

NOTE When the grade determined by gas content is different from that determined by gas pressure, the grade of higher risk shall be taken.

Explanation of Wording in This Specification

1. Words used for different degrees of strictness are explained as follows in order to mark the differences in executing the requirements in this specification.

    1) Words denoting a very strict or mandatory requirement:

        "Must" is used for affirmation; "must not" for negation.

    2) Words denoting a strict requirement under normal conditions:

        "Shall" is used for affirmation; "shall not" for negation.

    3) Words denoting a permission of a slight choice or an indication of the most suitable choice when conditions permit:

        "Should" is used for affirmation; "should not" for negation.

    4) "May" is used to express the option available, sometimes with the conditional permit.

2. "Shall meet the requirements of…" or "shall comply with…" is used in this specification to indicate that it is necessary to comply with the requirements stipulated in other relative standards and codes.

List of Quoted Standards

GB 50287,	Code for Hydropower Engineering Geological Investigation
GB 18871,	Basic Standards for Protection Against Ionizing Radiation and for the Safety of Radiation Sources
GBZ 116,	Standard for Controlling Radon and Its Progenies in Underground Space
NB/T 10074,	Specification for Engineering Geological Mapping of Hydropower Projects
NB/T 10075,	Specification for Karst Engineering Geological Investigation of Hydropower Projects
NB/T 10143,	Technical Code for Rockburst Risk Assessment of Hydropower Projects
NB/T 10227,	Code for Geophysical Exploration of Hydropower Projects
NB/T 10339,	Specification for Dam Site Engineering Geological Investigation of Hydropower Projects
NB/T 10340,	Specification for Pit Exploration of Hydropower Projects
NB/T 35007,	Geological Code of Construction Period for Hydropower Project
NB/T 35039,	Specification for Geological Observation of Hydropower Projects
NB/T 35052,	Specification of Water Quality Analysis for Hydropower Engineering Geological Investigation
NB/T 35098,	Specification of Regional Tectonic Stability Investigation for Hydropower Projects
NB/T 35102,	Specification for Soil Test in Situ in Borehole of Hydropower Projects
NB/T 35113,	Specification for Water Pressure Test in Borehole of Hydropower Projects
NB/T 35115,	Specification for Drilling Exploration of Hydropower Projects
DL/T 5006,	Code for Rock Mass Observations of Hydroelectric and Water Conservancy Engineering
DL/T 5367,	Code for Rock Mass Stress Measurements of Hydroelectric

and Water Conservancy Engineering

DL/T 5368, *Code for Rock Tests of Hydroelectric and Water Conservancy Engineering*